YOUR KNOWLEDGE HAS VALUE

- We will publish your bachelor's and master's thesis, essays and papers

- Your own eBook and book - sold worldwide in all relevant shops

- Earn money with each sale

Upload your text at www.GRIN.com
and publish for free

John Bredakis

Understanding the Zeta function, without getting lost in the tricky paths of advanced complex analysis

GRIN Publishing

Bibliographic information published by the German National Library:

The German National Library lists this publication in the National Bibliography; detailed bibliographic data are available on the Internet at http://dnb.dnb.de .

Imprint:

Copyright © 2013 GRIN Verlag, Open Publishing GmbH
Print and binding: Books on Demand GmbH, Norderstedt Germany
ISBN: 978-3-656-35430-7

This book at GRIN:

http://www.grin.com/en/e-book/207998/understanding-the-zeta-function-without-getting-lost-in-the-tricky-paths

GRIN - Your knowledge has value

Since its foundation in 1998, GRIN has specialized in publishing academic texts by students, college teachers and other academics as e-book and printed book. The website www.grin.com is an ideal platform for presenting term papers, final papers, scientific essays, dissertations and specialist books.

Visit us on the internet:

http://www.grin.com/

http://www.facebook.com/grincom

http://www.twitter.com/grin_com

The Zeta function for those outside the top club

of **Prime Numbers Theorem**

Realizing that the study of Zeta function $\zeta(s)$ is dependent on the Gamma

function $\Gamma(s)$, a function that I know well for $s=x \; \varepsilon \; R$, I decided to search

for the Zeta function in the internet ie: to get an overall satisfactory idea

about the Zeta function $\zeta(s) \quad s=(\sigma+i.t)$.

Analytic everywhere except at the single pole at s=1 , with residue one.

Fortunately the Gamma function is analytic in the complex plane with certain simple poles.

For example: $\lim s\to 0 \; s.\Gamma(s) = \lim s\to 0 \; \Gamma(s+1) = 0! =1$ Residue one

In my effort to search for $\zeta(s)$ I had to overcome my shalow knowledge

of complex analysis and my ignorance of the various delicate methods of

numerical analysis. (Intense study of very high mathematics , of very long duration).

As I mentioned my purpose was to get a satisfactory over all idea of the Zeta function $\zeta(s)$, without getting lost in the details of vast number of difficult to understand formulas.

From the initial frustration and dissapointment to the point of regretting my significant endeavour in higher mathematics,of very long duration ,to a final fullfilment of my purpose. **At least I believe so** !!

I relied mainly on the following sources:

1. Wikipedia the free encyclopedia   2. An indroduction to the Riemann hypothesis by Theodore Yoder and 3. The zeros on the critical line of the Zeta function by Lorenzo Menici.

http://Mathhighways.blogspot.com/
John Bredakis
jbredakis@gmail.com

Relation between the Zeta function and the primes

$$\zeta(s) = 1 + \frac{1}{2^s} + \frac{1}{3^s} + \frac{1}{4^s} + \frac{1}{5^s} + \cdots$$

$$\frac{1}{2^s}\zeta(s) = \frac{1}{2^s} + \frac{1}{4^s} + \frac{1}{6^s} + \frac{1}{8^s} + \frac{1}{10^s} + \cdots$$

Subtracting the second from the first we remove all elements that have a factor of 2:

$$\left(1 - \frac{1}{2^s}\right)\zeta(s) = 1 + \frac{1}{3^s} + \frac{1}{5^s} + \frac{1}{7^s} + \frac{1}{9^s} + \frac{1}{11^s} + \frac{1}{13^s} + \cdots$$

Repeating for the next term:

$$\frac{1}{3^s}\left(1 - \frac{1}{2^s}\right)\zeta(s) = \frac{1}{3^s} + \frac{1}{9^s} + \frac{1}{15^s} + \frac{1}{21^s} + \frac{1}{27^s} + \frac{1}{33^s} + \cdots$$

Subtracting again we get:

$$\left(1 - \frac{1}{3^s}\right)\left(1 - \frac{1}{2^s}\right)\zeta(s) = 1 + \frac{1}{5^s} + \frac{1}{7^s} + \frac{1}{11^s} + \frac{1}{13^s} + \frac{1}{17^s} + \cdots$$

where all elements having a factor of 3 or 2 (or both) are removed.

It can be seen that the right side is being sieved. Repeating infinitely we get:

$$\cdots \left(1 - \frac{1}{11^s}\right)\left(1 - \frac{1}{7^s}\right)\left(1 - \frac{1}{5^s}\right)\left(1 - \frac{1}{3^s}\right)\left(1 - \frac{1}{2^s}\right)\zeta(s) = 1$$

Dividing both sides by everything but the $\zeta(s)$ we obtain:

$$\zeta(s) = \frac{1}{\left(1 - \frac{1}{2^s}\right)\left(1 - \frac{1}{3^s}\right)\left(1 - \frac{1}{5^s}\right)\left(1 - \frac{1}{7^s}\right)\left(1 - \frac{1}{11^s}\right)\cdots}$$

General remarks on the Zeta function ζ(s)
And the first 42 roots in the critical strip

14.134725142	1	ζ(1) ->+oo
21.022039639	2	This sum can be expressed as:
25.010857580	3	a product of prime numbers ζ(3/2)=2.612
30.424876126	4	(See Wikipedia) 2
32.935061588	5	ζ(2)=π /6 =1.645
37.586178159	6	
40.918719012	7	+oo s ζ(3) =1.202
43.327073281	8	ζ(s) = Σ 1/n
48.005150881	9	n=1 4
49.773832478	10	ζ(4)=π /90 =1.0823
52.970321478		
56.446247697		
59.347044003		The zeta function in the complex plane has:
60.831778525		
65.112544048		
67.079810529		At s=1 a singularity:
69.546401711		(A single pole with residue one)
72.067157674		
75.704690699		At s=0 the value of -(1/2)
77.144840069	20	
79.337375020		
82.910380854		At s=-1 the value of -(1/12)
84.735492981		
87.425274613		
88.809111208		Trivial zeros at s = -2 , -4 , -6 , etc
92.491899271		
94.651344041		
95.870634228		The beauty of the zeta function is:
98.831194218		in the critical strip
101.317851006	30	0< Real s <1
103.725538040		with plenty of zeros
105.446623052		
107.168611184		The conjecture by Riemann is that:
111.029535543		all those zeros are in the line
111.874659177		Real s = (1/2)
114.320220915		
116.226680321		Notice that: ξ(1/2 + i.t) = ξ(1/2 - i.t)
118.790782866		
121.370125002		* Needless to say that the best
122.946829294	40	mathematical brains were
124.256818554	41	summond to develop
127.516683880	42	those supercomputer numbers

By Riemann Hypothesis : all the non trivial roots of the Zeta function ζ(s) in the critical strip (0<real s<1) are located on the critical line (real s=1/2)

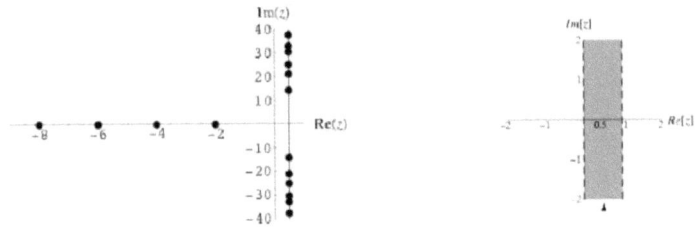

Understanding the Riemann Zeta function
And the Riemann Hypothesis

$$\Theta(x) = \sum_{n=-\infty}^{n=+\infty} e^{-n^2 . \pi . x} \qquad \frac{\Theta(x)-1}{2} = \sum_{n=1}^{+\infty} e^{-n^2 . \pi . x} \qquad \frac{\Theta(x)}{\Theta(1/x)} = \frac{1}{\sqrt{x}}$$

$$X(s) = \int_{0}^{+\infty} x^{s/2 - 1} . \left(\frac{\Theta(x)-1}{2}\right) . dx = \pi^{-s/2} . \Gamma\left(\frac{s}{2}\right) . \zeta(s)$$

$$X(s) = \frac{-1}{s.(1-s)} + \int_{1}^{+\infty} \left[x^{s/2 - 1} + x^{(1-s)/2 - 1} \right] . \left(\frac{\Theta(x)-1}{2}\right) . dx$$

$$\text{Real } s > 1$$

Notice that: $X(s) = X(1-s)$

$$X(s) = \pi^{-s/2} . \Gamma\left(\frac{s}{2}\right) . \zeta(s) = X(1-s) = \pi^{-(1-s)/2} . \Gamma\left(\frac{1-s}{2}\right) . \zeta(1-s)$$

Equivalent
formula $\qquad \zeta(s) = (2.\pi)^{s-1} . 2 . \sin\left(\frac{\pi.s}{2}\right) . \Gamma(1-s) . \zeta(1-s)$

Meaning that the formula of X(s)
is also valid for Real s<1

$$\zeta(s) = \frac{\pi^{s/2}}{\Gamma\left(\frac{s}{2}\right)} . X(s) = \frac{\pi^{s/2}}{\Gamma\left(\frac{s}{2}\right)} . \left[\frac{-1}{s.(1-s)} + \int_{1}^{+\infty} \left[x^{s/2} + x^{(1-s)/2} \right] . x^{-1} . \left(\frac{\Theta(x)-1}{2}\right) . dx \right]$$

$$\text{A convergent integral}$$

The above formula provides analytic continuation of $\zeta(s)$
and provides also the value of $\zeta(s)$ at s=0 ie: $\zeta(0)=-1/2$

$$\gamma = \text{Euler's constant} \qquad \frac{1}{\Gamma(s)} = e^{\gamma . s} . s . \prod_{n=1}^{+\infty} \left[1 + \frac{s}{n} \right] . e^{-s/n} \qquad \prod \text{ stands for products}$$

$\gamma = 0.577215665$

$$\theta(x) = \sum_{n=-\infty}^{n=+\infty} e^{-n^2 .\pi .x} \qquad \frac{\theta(x)-1}{2} = \sum_{n=1}^{+\infty} e^{-n^2 .\pi .x} \qquad \frac{\theta(x)}{\theta(1/x)} = \frac{1}{\sqrt{x}}$$

$$X(s) = \int_0^{+\infty} x^{s/2-1} .(\frac{\theta(x)-1}{2}).dx = \pi^{-s/2} .\Gamma(\frac{s}{2}).\zeta(s)$$

$$X(s) = \frac{-1}{s.(1-s)} + \int_1^{+\infty} \left[x^{s/2-1} + x^{(1-s)/2-1} \right] .(\frac{\theta(x)-1}{2}).dx$$
$$\text{Real } s > 1$$

$$X(s)=X(1-s) \qquad \textbf{Analytic Continuation}$$

$$\xi(s) = \frac{-s.(1-s)}{2} .X(s) = \frac{-s.(1-s)}{2} .\pi^{-s/2} .\Gamma(\frac{s}{2}).\zeta(s)$$

$$\xi(s) = \frac{1}{2} - \frac{s.(1-s)}{2} .\int_1^{+\infty} \left[x^{s/2-1} + x^{(1-s)/2-1} \right] .(\frac{\theta(x)-1}{2}).dx$$

$$\xi(\frac{1}{2} + i.t) = \frac{1}{2} - \frac{(1/4)+t^2}{2} .\int_1^{+\infty} x^{-3/4} .\cos\left[\ln(x).\frac{t}{2}\right] .(\theta(x)-1).dx$$

Notice that: $\xi(1/2 + i.t) = \xi(1/2 - i.t)$

For further details see also the pdf by Theodore Yoder
An Introduction to Riemann Hypothesis

The proof that:

(pdf by T.Yoder)

$$\xi\left(\frac{1}{2} + i.t\right) = \frac{1}{2} - \frac{(1/4)+t^2}{2} \cdot \int_{1}^{+\infty} x^{-3/4} \cdot \cos\left[\ln(x) \cdot \frac{t}{2}\right] \cdot (\Theta(x)-1) \cdot dx$$

$$\chi(s) := \int_{0}^{\infty} x^{s/2-1}\left(\frac{\vartheta(x)-1}{2}\right) dx = \int_{1}^{\infty} x^{s/2-1}\left(\frac{\vartheta(x)-1}{2}\right) dx$$
$$+ \int_{1}^{\infty} x^{-s/2-1}\left(\frac{\vartheta(1/x)-1}{2}\right) dx.$$

$$\frac{\vartheta(1/x)-1}{2} = \frac{\sqrt{x}\,\vartheta(x)-1}{2} = \sqrt{x}\left(\frac{\vartheta(x)-1}{2}\right) + \frac{\sqrt{x}-1}{2},$$

$$\chi(s) = \int_{1}^{\infty} (x^{s/2-1} + x^{(1-s)/2-1})\left(\frac{\vartheta(x)-1}{2}\right) dx$$
$$+ \int_{1}^{\infty} \frac{1}{2}\left[x^{(1-s)/2-1} - x^{-s/2-1}\right] dx.$$

Real s>1

$$\int_{1}^{\infty} \frac{1}{2}\left[x^{(1-s)/2-1} - x^{-s/2-1}\right] dx = \left[\frac{x^{(1-s)/2}}{1-s} + \frac{x^{-s/2}}{s}\right]_{1}^{\infty} = \frac{-1}{s(1-s)},$$

$$\chi(s) = \Gamma\left(\frac{s}{2}\right)\pi^{-s/2}\zeta(s)$$
$$= \frac{-1}{s(1-s)} + \int_{1}^{\infty} (x^{s/2-1} + x^{(1-s)/2-1})\left(\frac{\vartheta(x)-1}{2}\right) dx.$$

$$\frac{s(1-s)}{2} = \frac{(1/2+it)(1/2-it)}{2} = \frac{1/4+t^2}{2}.$$

s=(1/2+i.t)

$$(x^{s/2-1} + x^{(1-s)/2-1})\left(\frac{\vartheta(x)-1}{2}\right) = (x^{-3/4}x^{it/2} + x^{-3/4}x^{-it/2})\left(\frac{\vartheta(x)-1}{2}\right)$$
$$= x^{-3/4}\cos\left(\ln(x)\frac{t}{2}\right)(\vartheta(x)-1),$$

All the properties of the trigonometric functions can be derived by the relation

$$e^{i.p} = [\cos p + i.\sin p]$$

$$e^{i.p} = e^{\xi} = 1 + \xi + \frac{\xi^2}{2!} + \frac{\xi^3}{3!} + \frac{\xi^4}{4!} + \frac{\xi^5}{5!} + \dots$$

Convergent infinite series for any real p (! Factorial)

$$= [1 - p^2/2! + p^4/4! - ..] + i.[p - p^3/3! + p^5/5! - ...]$$

$$= \text{--------}\cos p\text{--------} + i. \text{--------}\sin p\text{--------}$$

Even powers-Even function   **Odd powers-Odd function**

Even: $f(x)=f(-x)$ Odd: $f(x)=-f(-x)$

$$e^{i.p}.e^{-i.p} = 1 = [\cos p + i.\sin p].[\cos p - i.\sin p] = [\cos^2 p + \sin^2 p]$$

$$e^{i.(p+q)} = [\cos(p+q) + i.\sin(p+q)]$$

$$e^{i.p}.e^{i.q} = [\cos p + i.\sin p].[\cos q + i.\sin q]$$

$$i^2 = -1$$

$\cos(p+q) = [\cos p.\cos q - \sin p.\sin q]$ Real part

$\sin(p+q) = [\sin p.\cos q + \cos p.\sin q]$ Imaginary part without the i in front

---

From
A List of MacLaurin series (Taylor's series for xi=0) $r_n(x) \to 0$

sinx	$x - [x^3/3!] + [x^5/5!] - [x^7/7!] +...$!=factorial	For any x
cosx	$1 - [x^2/2!] + [x^4/4!] - [x^6/6!] +...$	ie
e^x	$1 + x + [x^2/2!] + [x^3/3!] + [x^4/4!] +...$	$-\infty < x < +\infty$

By binomial expansion

$$e^x = \lim_{n \to +\infty} \left[1 + \frac{x}{n} \right]^n \quad \text{and} \quad e^{-x} = \lim_{n \to +\infty} \left[1 - \frac{x}{n} \right]^n$$

The way to handle the complex numbers
in the zeta function ζ(s) s=σ+**i**.t

$$\xi\left(\frac{1}{2}+it\right)=\frac{1}{2}\left[\frac{1}{2}+it\right]\left[-\frac{1}{2}+it\right]\Gamma\left(\frac{1}{4}+\frac{it}{2}\right)\pi^{-\frac{1}{4}-\frac{it}{2}}\zeta\left(\frac{1}{2}+it\right)$$

$$=-\frac{1}{2}\exp\left[\operatorname{Re}\log\Gamma\left(\frac{1}{4}+\frac{it}{2}\right)\right]\pi^{-\frac{1}{4}}\left(t^{2}+\frac{1}{4}\right)Z(t),$$

Riemann-Siegel Zeta-function   θ(0)=0   Z(0)=ζ(1/2)= - 1.46

$$Z(t)=e^{i\theta(t)}\zeta\left(\frac{1}{2}+it\right)=\exp\left[i\operatorname{Im}\log\Gamma\left(\frac{1}{4}+\frac{it}{2}\right)-i\frac{\log\pi}{2}t\right]\zeta\left(\frac{1}{2}+it\right)$$

For the above see next page

$$\Gamma(\frac{1}{4}+i.\frac{t}{2})=\Gamma(\Phi)=A+i.B=e^{\log[A+i.B]}$$

$$=e^{\operatorname{Re}\,\log\,\Gamma(\Phi)}\cdot e^{i.\operatorname{Im}\,\log\,\Gamma(\Phi)}\qquad \Phi=(\frac{1}{4}+i.\frac{t}{2})$$

$$=\exp\left[\operatorname{Re}\,\log\,\Gamma(\Phi)\right]\cdot\exp\left[i.\operatorname{Im}\,\log\,\Gamma(\Phi)\right]$$

$$x^{i.(\frac{t}{2})}=\left[e^{\log x}\right]^{i.(\frac{t}{2})}=e^{i.\log(x).(\frac{t}{2})}$$

$$=\cos\left[\log(x).(\frac{t}{2})\right]+i.\sin\left[\log(x).(\frac{t}{2})\right]$$

- -

$$x^{i.(\frac{-t}{2})}=\left[e^{\log x}\right]^{i.(\frac{-t}{2})}=e^{i.\log(x).(\frac{-t}{2})}$$

$$=\cos\left[\log(x).(\frac{-t}{2})\right]+i.\sin\left[\log(x).(\frac{-t}{2})\right]$$

$$=\cos\left[\log(x).(\frac{t}{2})\right]-i.\sin\left[\log(x).(\frac{t}{2})\right]$$

$$\frac{x^{i.\frac{t}{2}}+x^{-i.\frac{t}{2}}}{2}=\cos\left[\log(x).\frac{t}{2}\right]$$

Summary of the Riemann Hypothesis

$$\theta(x) = \sum_{n=-\infty}^{n=+\infty} e^{-n^2 \cdot \pi \cdot x} \qquad \frac{\theta(x)-1}{2} = \sum_{n=1}^{+\infty} e^{-n^2 \cdot \pi \cdot x} \qquad \frac{\theta(x)}{\theta(1/x)} = \frac{1}{\sqrt{x}}$$

$$X(s) = \int_0^{+\infty} x^{s/2-1} \cdot \left(\frac{\theta(x)-1}{2}\right) \cdot dx = \pi^{-s/2} \cdot \Gamma\left(\frac{s}{2}\right) \cdot \zeta(s)$$

$$X(s) = \frac{-1}{s \cdot (1-s)} + \int_1^{+\infty} \left[x^{s/2-1} + x^{(1-s)/2-1} \right] \cdot \left(\frac{\theta(x)-1}{2}\right) \cdot dx$$

Real s > 1

X(s)=X(1-s)   **Analytic continuation**
See: Understanding the Zeta function

$$\xi(s) = \frac{-s \cdot (1-s)}{2} \cdot X(s) = \frac{-s \cdot (1-s)}{2} \cdot \pi^{-s/2} \cdot \Gamma\left(\frac{s}{2}\right) \cdot \zeta(s)$$

$$\xi(s) = \frac{1}{2} - \frac{s \cdot (1-s)}{2} \cdot \int_1^{+\infty} \left[x^{s/2-1} + x^{(1-s)/2-1} \right] \cdot \left(\frac{\theta(x)-1}{2}\right) \cdot dx$$

$$\xi\left(\frac{1}{2} + i.t\right) = \frac{1}{2} - \frac{(1/4)+t^2}{2} \cdot \int_1^{+\infty} x^{-3/4} \cdot \cos\left[\ln(x) \cdot \frac{t}{2}\right] \cdot (\theta(x)-1) \cdot dx$$

$$\xi(1/2 + i.t) = \xi(1/2 - i.t)$$

$$\xi\left(\frac{1}{2}+it\right) = \frac{1}{2}\left[\frac{1}{2}+it\right]\left[-\frac{1}{2}+it\right]\Gamma\left(\frac{1}{4}+\frac{it}{2}\right)\pi^{-\frac{1}{4}-\frac{it}{2}}\zeta\left(\frac{1}{2}+it\right)$$

$$= -\frac{1}{2}\exp\left[\operatorname{Re}\log\Gamma\left(\frac{1}{4}+\frac{it}{2}\right)\right]\pi^{-\frac{1}{4}}\left(t^2+\frac{1}{4}\right)Z(t),$$

Riemann-Siegel Zeta-function   θ(0)=0   Z(0)=ζ(1/2)= - 1.46

$$Z(t) = e^{i\vartheta(t)}\zeta\left(\frac{1}{2}+it\right) = \exp\left[i\operatorname{Im}\log\Gamma\left(\frac{1}{4}+\frac{it}{2}\right) - i\frac{\log\pi}{2}t\right]\zeta\left(\frac{1}{2}+it\right)$$

$$\exp\left[\operatorname{Re}\,\log\,\Gamma\left(\frac{1}{4}+\frac{i.t}{2}\right)\right] = \exp\left[\operatorname{Re}\,\log\,\Gamma\left(\frac{1}{4}-\frac{i.t}{2}\right)\right]$$

$$z(t) = e^{i.\theta(t)} \cdot \zeta\left(\frac{1}{2}+i.t\right) = e^{-i.\theta(t)} \cdot \zeta\left(\frac{1}{2}-i.t\right)$$

Graph of Z(t) in the range of t (0⩽ t ⩽50)

$$Z(t) = e^{i\theta(t)} \sum_{n=1}^{[x]} \frac{1}{n^{\frac{1}{2}+it}} + e^{-i\theta(t)} \sum_{n=1}^{[x]} \frac{1}{n^{\frac{1}{2}-it}} + O(t^{-\frac{1}{4}})$$

$$= 2 \sum_{n=1}^{[x]} \frac{\cos(\theta(t) - t\log n)}{n^{\frac{1}{2}}} + O(t^{-\frac{1}{4}}).$$

[x]=m

$$\tau = \sqrt{\frac{t}{2\pi}}, \quad m = \lfloor \tau \rfloor, \quad z = 2(t-m) - 1$$

**The estimation of the remaining terms is a science by itself
exclusively for the members of the top club
of Prime Numbers Theorem**

$$Z(t) = 2 \sum_{n=1}^{m} \frac{\cos(\theta(t) - t\log n)}{\sqrt{n}} +$$

$$(-1)^{m+1}\tau^{-1/2} \sum_{j=0}^{M} (-1)^j \tau^{-j} \Phi_j(z) + R_M(t)$$

The first functions $\Phi_j(z)$ are defined by

$$\Phi_0(z) = \frac{\cos(\frac{1}{2}\pi z^2 + \frac{3}{8}\pi)}{\cos(\pi z)}$$

$$\Phi_1(z) = \frac{1}{12\pi^2} \Phi_0^{(3)}(z)$$

$$\Phi_2(z) = \frac{1}{16\pi^2} \Phi_0^{(2)}(z) + \frac{1}{288\pi^4} \Phi_0^{(6)}(z)$$

The general expression of $\Phi_j(z)$ for $j > 2$ is quite complicated and we refer to [6]
or [10] for it. As exposed in [2], explicit bounds have been rigorously obtained
on the error term $R_M(t)$, and for $t \geq 200$, one has

$$|R_0(t)| \leq 0.127\, t^{-3/4}, \quad |R_1(t)| \leq 0.053\, t^{-5/4}, \quad |R_2(t)| \leq 0.011\, t^{-7/4}.$$

Numerical evaluation of the Riemann Zeta function
Xavier Gourdon and Pascal Sebah
July 23, 2003[1]

To be able to completly approximate $Z(t)$ thanks to formula (9), it remains to give an approximation of the $\theta(t)$ function which is obtained from expression (6) using Stirling series, giving

$$\theta(t) = \frac{t}{2} \log \frac{t}{2\pi} - \frac{t}{2} - \frac{\pi}{8} + \frac{1}{48t} + \frac{7}{5760t^3} + \cdots \qquad (10)$$

Practical approximation considerations

For practical purposes in computations relative to zeros of $\zeta(s)$ it is not necessary to compute precisely the zeros but just to locate them, and using $M = 1$ or $M = 2$ in formula (9) is usually sufficient. For t around 10^{10} for example, the choice $M = 1$ permits to obtain an absolute precision of $Z(t)$ smaller than 2×10^{-14}, and with $M = 2$ the precision is smaller than 5×10^{-20}. As for the

number of terms involved in the sum of (9), it is proportional to $\sqrt{t}$ which is much better than previous approaches without Riemann-Siegel formula which required a number of terms of order t. For $t \simeq 10^{10}$ for example, Riemann-Siegel formula only requires $m \simeq 40,000$ terms, whereas approach of proposition 1 requires at least $\simeq 9 \times 10^9$ terms.

Fractional part

All real numbers can be written in the form $n + r$ where n is an integer (the integer part) and the remaining **fractional part** r is a nonnegative real number less than one. For a positive number written in decimal notation, the fractional part corresponds to the digits appearing after the decimal point.

The fractional part of a real number x is $x - \lfloor x \rfloor$, where $\lfloor \ \rfloor$ is the floor function. It is sometimes denoted $\{x\}$, $\langle x \rangle$ or $x \bmod 1$.

$$Z(t) = e^{i\theta(t)} \sum_{n=1}^{[x]} \frac{1}{n^{\frac{1}{2}+it}} + e^{-i\theta(t)} \sum_{n=1}^{[x]} \frac{1}{n^{\frac{1}{2}-it}} + O(t^{-\frac{1}{4}})$$

$$= 2 \sum_{n=1}^{[x]} \frac{\cos(\theta(t) - t \log n)}{n^{\frac{1}{2}}} + O(t^{-\frac{1}{4}}).$$

[x]=m

$$\tau = \sqrt{\frac{t}{2\pi}}, \quad m = \lfloor \tau \rfloor, \quad z = 2(t-m) - 1$$

Explanation:

$$\zeta(s) = (2\pi)^{s-1} . 2 . \sin(\frac{\pi . s}{2}) . \Gamma(1-s) . \zeta(1-s) = \frac{\pi^{-(1-s)/2} . \Gamma(\frac{1-s}{2})}{\pi^{-s/2} . \Gamma(\frac{s}{2})} . \zeta(1-s)$$

--------=X(s)-----------

*Terminology by Gourdon-Sebah

$$\zeta(\frac{1}{2} + i.t) = \frac{\pi^{i.t} . \Gamma(\frac{1}{4} - i.\xi)}{\Gamma(\frac{1}{4} + i.\xi)} . \zeta(\frac{1}{2} - i.t) = X(\frac{1}{2} + i.t) . \zeta(\frac{1}{2} - i.t)$$

$\xi = t/2$

Notice that: $X(\frac{1}{2} + i.t) . X(\frac{1}{2} - i.t) = 1$

$$e^{i.\theta(t)} = \frac{1}{\left[X(\frac{1}{2} + i.t)\right]^{1/2}}$$

$$Z(t) = e^{i.\theta(t)} . \zeta(\frac{1}{2} + i.t) = \left[X(\frac{1}{2} + i.t)\right]^{-1/2} . \left[X(\frac{1}{2} + i.t)\right] . \zeta(\frac{1}{2} - i.t)$$

$$= \left[X(\frac{1}{2} + i.t)\right]^{1/2} . \zeta(\frac{1}{2} - i.t)$$

$$= e^{-i.\theta(t)} . \zeta(\frac{1}{2} - i.t)$$

$$\vartheta(t) \sim \frac{t}{2} \log \frac{t}{2\pi} - \frac{t}{2} - \frac{\pi}{8} + \frac{1}{48t}$$

$\theta(0) = 0$

Devoted to Uncle Fotis , my Mentor in Mathematics

The easiest proof of the following:

$$Z(t) = e^{i.\theta(t)} . \zeta(\frac{1}{2} + i.t) = e^{-i.\theta(t)} . \zeta(\frac{1}{2} - i.t)$$

$$\xi(s) = \frac{-s.(1-s)}{2} . X(s) = \frac{-s.(1-s)}{2} . \pi^{-s/2} . \Gamma(\frac{s}{2}) . \zeta(s)$$

$$\xi(s) = \frac{1}{2} - \frac{s.(1-s)}{2} . \int_1^{+oo} [x^{s/2 - 1} + x^{(1-s)/2 - 1}] . (\frac{\theta(x)-1}{2}) . dx$$

$$\xi(\frac{1}{2} + i.t) = \frac{1}{2} - \frac{(1/4)+t^2}{2} . \int_1^{+oo} x^{-3/4} . \cos[\ln(x).\frac{t}{2}] . (\theta(x)-1) . dx$$

$$\xi(1/2 + i.t) = \xi(1/2 - i.t)$$

$$\xi\left(\frac{1}{2} + it\right) = \frac{1}{2}\left[\frac{1}{2} + it\right]\left[-\frac{1}{2} + it\right]\Gamma\left(\frac{1}{4} + \frac{it}{2}\right)\pi^{-\frac{1}{4}-\frac{it}{2}}\zeta\left(\frac{1}{2} + it\right)$$

$$= -\frac{1}{2}\exp\left[\operatorname{Re}\log\Gamma\left(\frac{1}{4} + \frac{it}{2}\right)\right]\pi^{-\frac{1}{4}}\left(t^2 + \frac{1}{4}\right)Z(t) ,$$

$$\theta(0)=0 \quad \zeta(1/2)=-1.46$$

$$Z(t) = e^{i\theta(t)}\zeta\left(\frac{1}{2} + it\right) = \exp\left[i\operatorname{Im}\log\Gamma\left(\frac{1}{4} + \frac{it}{2}\right) - i\frac{\log\pi}{2}t\right]\zeta\left(\frac{1}{2} + it\right)$$

Z(t) is the Riemann Siegel function

$$\vartheta(t) \sim \frac{t}{2}\log\frac{t}{2\pi} - \frac{t}{2} - \frac{\pi}{8} + \frac{1}{48t}$$

$$\exp\left[\operatorname{Re}\,\log\,\Gamma(-\frac{1}{4} + \frac{i.t}{2})\right] = \exp\left[\operatorname{Re}\,\log\,\Gamma(-\frac{1}{4} - \frac{i.t}{2})\right]$$

Haselgrove's table for $\zeta(1/2+i.t)$ in the range $0\le t \le 26.8$

t	$\zeta(1/2+i.t)$	t	$\zeta(1/2+i.t)$	t	$\zeta(1/2+i.t)$
0.0	-1.46	9.0	+1.45+0.19.i	18.0	+2.33-0.19.i
0.2	-1.18-0.67.i	9.2	+1.48+0.14.i	18.2	+2.27-0.45.i
0.4	-0.68-0.94.i	9.4	+1.51+0.08.i	18.4	+2.17-0.66.i
0.6	-0.28-0.94.i	9.6	+1.53+0.02.i	18.6	+2.02-0.86.i
0.8	-0.02-0.84.i	9.8	+1.54-0.04.i	18.8	+1.84-1.03.i
1.0	+0.14-0.72.i	10.0	+1.54-0.12.i	19.0	+1.62-1.16.i
1.2	+0.25-0.62.i	10.2	+1.54-0.19.i	19.2	+1.38-1.24.i
1.4	+0.32-0.52.i	10.4	+1.53-0.26.i	19.4	+1.13-1.28.i
1.6	+0.37-0.44.i	10.6	+1.50-0.34.i	19.6	+0.88-1.26.i
1.8	+0.41-0.37.i	10.8	+1.47-0.42.i	19.8	+0.65-1.18.i
2.0	+0.44-0.31.i	11.0	+1.42-0.49.i	20.0	+0.43-1.06.i
2.2	+0.46-0.26.i	11.2	+1.36-0.56.i	20.2	+0.25-0.90.i
2.4	+0.48-0.21.i	11.4	+1.29-0.62.i	20.4	+0.11-0.70.i
2.6	+0.50-0.16.i	11.6	+1.21-0.67.i	20.6	+0.02-0.48.i
2.8	+0.52-0.12.i	11.8	+1.12-0.71.i	20.8	-0.02-0.25.i
3.0	+0.53-0.08.i	12.0	+1.02-0.75.i	21.0	-0.01-0.02.i
3.2	+0.55-0.04.i	12.2	+0.91-0.76.i	21.2	+0.06+0.19.i
3.4	+0.56-0.01.i	12.4	+0.79-0.76.i	21.4	+0.18+0.35.i
3.6	+0.58+0.03.i	12.6	+0.68-0.75.i	21.6	+0.34+0.52.i
3.8	+0.59+0.06.i	12.8	+0.56-0.71.i	21.8	+0.52+0.62.i
4.0	+0.61+0.05.i	13.0	+0.44-0.06.i	22.0	+0.72+0.67.i
4.2	+0.62+0.12.i	13.2	+0.33-0.58.i	22.2	+0.92+0.66.i
4.4	+0.64+0.15.i	13.4	+0.23-0.49.i	22.4	+1.11+0.60.i
4.6	+0.66+0.18.i	13.6	+0.35-0.38.i	22.6	+1.26+0.49.i
4.8	+0.68+0.21.i	13.8	+0.07-0.24.i	22.8	+1.38+0.34.i
5.0	+0.70+0.23.i	14.0	+0.02-0.10.i	23.0	+1.45+0.16.i
5.2	+0.73+0.26.i	14.2	-0.01+0.04.1	23.2	+1.46-0.03.i
5.4	+0.75+0.28.i	14.4	-0.01+0.21.i	23.4	+1.41-0.21.i
5.6	+0.78+0.30.i	14.6	+0.01+0.38.i	23.6	+1.30-0.38.i
5.8	+0.81+0.32.i	14.8	+0.07+0.55.i	23.8	+1.14-0.50.i
6.0	+0.84+0.34.i	15.0	+0.15+0.70.i	24.0	+0.95-0.58.i
6.2	+0.87+0.36.i	15.2	+0.26+0.85.i	24.2	+0.73-0.60.i
6.4	+0.91+0.37.i	15.4	+0.39+0.98.i	24.4	+0.51-0.55.i
6.6	+0.94+0.38.i	15.6	+0.56+1.09.i	24.6	+0.30-0.43.i
6.8	+0.98+0.39.i	15.8	+0.74+1.17.1	24.8	+0.13-0.21.i
7.0	+1.02+0.40.i	16.0	+0.94+1.22.i	25.0	0.00-0.01.i
7.2	+1.06+0.40.i	16.2	+1.15+1.23.i	25.2	-0.05+0.26.i
7.4	+1.11+0.40.i	16.4	+1.36+1.20.i	25.4	-0.04+0.55.i
7.6	+1.15+0.39.i	16.6	+1.57+1.14.i	25.6	+0.06+0.85.i
7.8	+1.20+0.38.i	16.8	+1.77+1.04.i	25.8	+0.25+1.11.i
8.0	+1.24+0.36.i	17.0	+1.95+0.90.i	26.0	+0.50+1.34.i
8.2	+1.29+0.34.i	17.2	+2.10+0.72.i	26.2	+0.82+1.49.i
8.4	+1.33+0.31.i	17.4	+2.22+0.52.i	26.4	+1.17+1.56.i
8.6	+1.37+0.28.i	17.6	+2.30+0.29.i	26.8	+1.55+1.54.i
8.8	+1.41+0.24.i	17.8	+2.34+0.06.i	26.8	+1.92+1.42.i

Roots of $\zeta(1/2+i.t)$ in the range $0\le t \le 26.8$

14.134725142
21.022039639
25.010857580

Haselgrove's table for $\zeta(\tfrac{1}{2} + it)$ in the range $27.0 \le t \le 50.0$.

t	$\zeta(\tfrac{1}{2} + it)$	t	$\zeta(\tfrac{1}{2} + it)$	t	$\zeta(\tfrac{1}{2} + it)$
27.0	+2.25 + 1.21i	35.0	+2.60 + 1.11i	43.0	+0.44 − 0.31i
27.2	+2.53 + 0.91i	35.2	+2.84 + 0.67i	43.2	+0.16 − 0.16i
27.4	+2.73 + 0.55i	35.4	+2.94 + 0.17i	43.4	−0.07 + 0.11i
27.6	+2.83 + 0.15i	35.6	+2.89 − 0.33i	43.6	−0.20 + 0.50i
27.8	+2.83 − 0.27i	35.8	+2.70 − 0.80i	43.8	−0.18 + 0.94i
28.0	+2.72 − 0.68i	36.0	+2.38 − 1.19i	44.0	+0.01 + 1.40i
28.2	+2.52 − 1.05i	36.2	+1.97 − 1.46i	44.2	+0.37 + 1.80i
28.4	+2.23 − 1.35i	36.4	+1.50 − 1.59i	44.4	+0.87 + 2.08i
28.6	+1.87 − 1.57i	36.6	+1.03 − 1.57i	44.6	+1.47 + 2.19i
28.8	+1.48 − 1.69i	36.8	+0.60 − 1.40i	44.8	+2.11 + 2.10i
29.0	+1.09 − 1.70i	37.0	+0.26 − 1.12i	45.0	+2.71 + 1.80i
29.2	+0.71 − 1.61i	37.2	+0.04 − 0.76i	45.2	+3.21 + 1.31i
29.4	+0.38 − 1.43i	37.4	−0.05 − 0.36i	45.4	+3.54 + 0.69i
29.6	+0.13 − 1.18i	37.6	+0.01 + 0.03i	45.6	+3.66 − 0.03i
29.8	−0.04 − 0.89i	37.8	+0.19 + 0.36i	45.8	+3.56 − 0.74i
30.0	−0.12 − 0.58i	38.0	+0.46 + 0.59i	46.0	+3.24 − 1.39i
30.2	−0.11 − 0.29i	38.2	+0.80 + 0.71i	46.2	+2.75 − 1.90i
30.4	−0.02 − 0.03i	38.4	+1.14 + 0.69i	46.4	+2.14 − 2.22i
30.6	+0.14 + 0.17i	38.6	+1.44 + 0.55i	46.6	+1.49 − 2.33i
30.8	+0.33 + 0.30i	38.8	+1.67 + 0.31i	46.8	+0.86 − 2.24i
31.0	+0.52 + 0.34i	39.0	+1.79 ± 0.00i	47.0	+0.33 − 1.97i
31.2	+0.70 + 0.31i	39.2	+1.78 − 0.33i	47.2	−0.06 − 1.57i
31.4	+0.84 + 0.22i	39.4	+1.66 − 0.64i	47.4	−0.27 − 1.11i
31.6	+0.92 + 0.09i	39.6	+1.43 − 0.88i	47.6	−0.31 − 0.66i
31.8	+0.92 − 0.06i	39.8	+1.12 − 1.02i	47.8	−0.21 − 0.28i
32.0	+0.84 − 0.20i	40.0	+0.79 − 1.04i	48.0	−0.01 − 0.01i
32.2	+0.71 − 0.29i	40.2	+0.48 − 0.95i	48.2	+0.24 + 0.14i
32.4	+0.52 − 0.32i	40.4	+0.22 − 0.75i	48.4	+0.47 + 0.15i
32.6	+0.31 − 0.27i	40.6	+0.05 − 0.48i	48.6	+0.64 + 0.07i
32.8	+0.11 − 0.14i	40.8	−0.02 − 0.18i	48.8	+0.71 − 0.06i
33.0	−0.05 + 0.08i	41.0	+0.03 + 0.12i	49.0	+0.67 − 0.20i
33.2	−0.13 + 0.36i	41.2	+0.18 + 0.36i	49.2	+0.53 − 0.29i
33.4	−0.13 + 0.69i	41.4	+0.40 + 0.51i	49.4	+0.34 − 0.29i
33.6	−0.02 + 1.03i	41.6	+0.64 + 0.57i	49.6	+0.14 − 0.18i
33.8	+0.20 + 1.35i	41.8	+0.87 + 0.52i	49.8	−0.02 + 0.03i
34.0	+0.52 + 1.60i	42.0	+1.04 + 0.37i	50.0	−0.08 + 0.33i
34.2	+0.92 + 1.75i	42.2	+1.12 + 0.18i		
34.4	+1.37 + 1.79i	42.4	+1.08 − 0.04i		
34.6	+1.83 + 1.69i	42.6	+0.95 − 0.22i		
34.8	+2.25 + 1.46i	42.8	+0.72 − 0.32i		

Roots in the above range: 30.424876126 32.935061588

37.58617815 40.918719012 43.327073281

48.005150881 49.773832478

Remarks about analytic continuation

The formula of X(s) is derived by the assumption that Real s>1

$$\Theta(x) = \sum_{n=-\infty}^{n=+\infty} e^{-n^2.\pi.x} \qquad \left| \quad \frac{\Theta(x)-1}{2} = \sum_{n=1}^{+\infty} e^{-n^2.\pi.x} \quad \right| \quad \frac{\Theta(x)}{\Theta(1/x)} = \frac{1}{\sqrt{x}}$$

$$X(s) = \int_0^{+\infty} x^{s/2 - 1} . \left(\frac{\Theta(x)-1}{2}\right).dx = \pi^{-s/2} . \Gamma(\frac{s}{2}).\zeta(s)$$

$$X(s) = \frac{-1}{s.(1-s)} + \int_1^{+\infty} \left[x^{s/2 - 1} + x^{(1-s)/2 - 1} \right] . \left(\frac{\Theta(x)-1}{2}\right).dx$$

$$\text{Real } s > 1$$

Notice that: X(s) = X(1-s)

$$X(s) = \pi^{-s/2} . \Gamma(\frac{s}{2}).\zeta(s) = X(1-s) = \pi^{-(1-s)/2} . \Gamma(\frac{1-s}{2}).\zeta(1-s)$$

Equivalent
formula $\qquad \zeta(s) = (2.\pi)^{s-1} . 2 . \sin(\frac{\pi.s}{2}).\Gamma(1-s).\zeta(1-s)$

Meaning that the formula of X(s)
is also valid for Real s<1

$$\zeta(s) = \frac{\pi^{s/2}}{\Gamma(\frac{s}{2})}.X(s) = \frac{\pi^{s/2}}{\Gamma(\frac{s}{2})} . \left| \frac{-1}{s.(1-s)} + \int_1^{+\infty} \left[x^{s/2} + x^{(1-s)/2} \right] . x^{-1} . \left(\frac{\Theta(x)-1}{2}\right).dx \right|$$

$$\text{A convergent integral}$$

The above formula provides analytic continuation of $\zeta(s)$
and provides also the value of $\zeta(s)$ at s=0 ie: $\zeta(0)=-1/2$

γ=Euler's constant
|
=0.577215665

$$\frac{1}{\Gamma(s)} = e^{\gamma.s} . s . \prod_{n=1}^{+\infty} \left[1 + \frac{s}{n} \right] . e^{-s/n} \qquad \prod \text{ stands for products}$$

By analytic continuation of $\zeta(s)$
except for the pole $(s=1)$

This formula is analytic also in the critical strip
$$0 < \text{Real } s < 1$$

$$\zeta(s) = \frac{\pi^{s/2}}{\Gamma(\frac{s}{2})} \cdot X(s) = \frac{\pi^{s/2}}{\Gamma(\frac{s}{2})} \cdot \left[\frac{-1}{s.(1-s)} + \int_1^{+\infty} \left[x^{s/2} + x^{(1-s)/2} \right] . x^{-1} . \left(\frac{\Theta(x)-1}{2}\right) . dx \right]$$

A convergent integral

Therefore the following formulas are also analytic in the critical strip

$$\xi(s) = \frac{-s.(1-s)}{2} . X(s) = \frac{-s.(1-s)}{2} . \pi^{-s/2} . \Gamma(\frac{s}{2}) . \zeta(s)$$

$$\xi(s) = \frac{1}{2} - \frac{s.(1-s)}{2} . \int_1^{+\infty} \left[x^{s/2-1} + x^{(1-s)/2-1} \right] . \left(\frac{\Theta(x)-1}{2}\right) . dx$$

$$\xi(\frac{1}{2} + i.t) = \frac{1}{2} - \frac{(1/4)+t^2}{2} . \int_1^{+\infty} x^{-3/4} . \cos\left[\ln(x) . \frac{t}{2}\right] . (\Theta(x)-1) . dx$$

The tricks are

1. The analytic continuation of $\zeta(s)$
   With $\zeta(0) = -1/2$

and

2. The convergence of the integral

$$\zeta(s) = \frac{\pi^{s/2}}{\Gamma(\frac{s}{2})} . X(s) = \frac{\pi^{s/2}}{\Gamma(\frac{s}{2})} . \left[\frac{-1}{s.(1-s)} + \int_1^{+\infty} \left[x^{s/2} + x^{(1-s)/2} \right] . x^{-1} . \left(\frac{\Theta(x)-1}{2}\right) . dx \right]$$

A convergent integral

$$\Theta(x) = \sum_{n=-\infty}^{n=+\infty} e^{-n^2 . \pi . x} \quad \left| \quad \frac{\Theta(x)-1}{2} = \sum_{n=1}^{+\infty} e^{-n^2 . \pi . x} \quad \right| \quad \frac{\Theta(x)}{\Theta(1/x)} = \frac{1}{\sqrt{x}}$$

See also the pdf by Lorenzo Menici
for the complete proof of the above

Notice also:

$$\zeta(s) = \frac{\pi^{s/2}}{\Gamma(\frac{s}{2})}.X(s) = \frac{1}{\Gamma(s)}.\int_0^{+\infty} \frac{1}{(e^t - 1)}.t^{s-1}.dt = \frac{1}{\Gamma(s)}.\sum_{n=1}^{+\infty}.\int_0^{+\infty} e^{-n.t}.M$$

$$M = t^{s-1}.dt$$

$$\boxed{0 < \text{Real } s < 1} = \frac{1}{\Gamma(s)}.\int_0^{+\infty} \left[\frac{1}{(e^t - 1)} - \frac{1}{t} \right].t^{s-1}.dt \quad \begin{array}{c} * \\ \text{pdf on zeta} \\ \text{function} \\ \textbf{by Petersen} \end{array}$$

$$\boxed{-1 < \text{Real } s < 0} = \frac{1}{\Gamma(s)}.\int_0^{+\infty} \left[\frac{1}{(e^t - 1)} - \frac{1}{t} + \frac{1}{2} \right].t^{s-1}.dt \quad *$$

The Zeta function in the complex plain

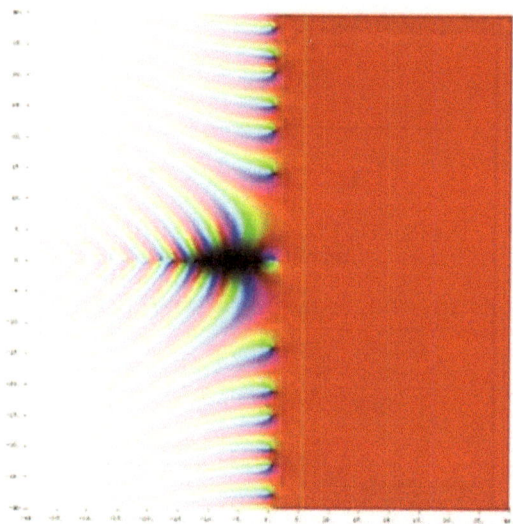

Some additional remarks on the Zeta function

An observation: In general:

$$\frac{10}{\ln 10} \sim 4 = \pi(10) =$$

| Number of primes |
| up to the number 10 |

$$\pi(x) \sim \frac{x}{\ln x}$$

        2 3 5 7   As x->+oo

Bernoulli's numbers

$$B2 = \frac{1}{6}$$

$$\zeta(-n) = (-1)^n \cdot \frac{Bn+1}{n+1}$$

$$B4 = \frac{-1}{30}$$

| Odd Bernoulli's |
| numbers |
| with the exception |
| of B1 vanish |

$$B6 = \frac{1}{42}$$

$$B8 = \frac{-1}{30}$$

$$\zeta(2n) = \frac{(2\pi)^{2n} \cdot (-1)^{n+1} \cdot B2n}{2 \cdot (2n)!}$$

- -

$$\zeta(s) = \left[1 - \frac{2}{2^s}\right]^{-1} \cdot \sum_{n=1}^{+oo} \frac{(-1)^{n+1}}{n^s}$$

Converges uniformly for Real s>0

See also the pdf by Theodore Yoder
for repulsion between the zeros of $\zeta(s)$
in the critical line

Odlyzko–Schönhage algorithm

**Exclusively for the members of the top club
of Prime Numbers Theorem**

$$Z(t) = 2 \sum_{n \le \sqrt{t/(2\pi)}} \frac{\cos\left[\vartheta(t) - t\log n\right]}{\sqrt{n}} + r(t) \tag{2.34}$$

is approximately

$$r(t) \sim (-1)^{N-1} \left(\frac{t}{2\pi}\right)^{-\frac{1}{4}} \left[A_0 + A_1 \left(\frac{t}{2\pi}\right)^{-\frac{1}{2}} + A_2 \left(\frac{t}{2\pi}\right)^{-1} + A_3 \left(\frac{t}{2\pi}\right)^{-\frac{3}{2}} + A_4 \left(\frac{t}{2\pi}\right)^{-2} \right] \tag{2.35}$$

where $N = [\sqrt{t/(2\pi)}]$, $p = \{\sqrt{t/(2\pi)}\}$ *and*

$$A_0 = \Psi(p) = \frac{\cos\left[2\pi(p^2 - p - \frac{1}{16})\right]}{\cos(2\pi p)},$$

$$A_1 = -\frac{1}{2^5 \, 3\pi^2} \Psi^{(3)}(p),$$

$$A_2 = \frac{1}{2^6 \pi^2} \Psi^{(2)}(p) + \frac{1}{2^{11} \, 3^2 \pi^4} \Psi^{(6)}(p),$$

$$A_3 = -\frac{1}{2^6 \pi^2} \Psi^{(1)}(p) - \frac{1}{2^8 \, 3 \cdot 5\pi^2} \Psi^{(5)}(p) - \frac{1}{2^{16} \, 3^4 \pi^6} \Psi^{(8)}(p),$$

$$A_4 = \frac{1}{2^7 \pi^2} \Psi(p) + \frac{1}{2^{13} \, 3\pi^4} \Psi^{(4)}(p) + \frac{1}{2^{17} \, 3^2 \cdot 5\pi^6} \Psi^{(8)}(p) + \frac{1}{2^{23} \, 3^5 \pi^8} \Psi^{(12)}(p).$$

See also Haselgrove's tables

Summary of the Riemann Hypothesis

$$\theta(x) = \sum_{n=-\infty}^{n=+\infty} e^{-n^2 . \pi . x} \qquad \frac{\theta(x)-1}{2} = \sum_{n=1}^{+\infty} e^{-n^2 . \pi . x} \qquad \frac{\theta(x)}{\theta(1/x)} = \frac{1}{\sqrt{x}}$$

$$X(s) = \int_0^{+\infty} x^{s/2 - 1} . \left(\frac{\theta(x)-1}{2}\right) . dx = \pi^{-s/2} . \Gamma\left(\frac{s}{2}\right) . \zeta(s)$$

$$X(s) = \frac{-1}{s.(1-s)} + \int_1^{+\infty} \left[x^{s/2 - 1} + x^{(1-s)/2 - 1} \right] . \left(\frac{\theta(x)-1}{2}\right) . dx$$

Real s > 1

X(s)=X(1-s)   **Analytic continuation**

See: Understanding the Zeta function

$$\xi(s) = \frac{-s.(1-s)}{2} . X(s) = \frac{-s.(1-s)}{2} . \pi^{-s/2} . \Gamma\left(\frac{s}{2}\right) . \zeta(s)$$

$$\xi(s) = \frac{1}{2} - \frac{s.(1-s)}{2} . \int_1^{+\infty} \left[x^{s/2 - 1} + x^{(1-s)/2 - 1} \right] . \left(\frac{\theta(x)-1}{2}\right) . dx$$

$$\xi\left(\frac{1}{2} + i.t\right) = \frac{1}{2} - \frac{(1/4)+t^2}{2} . \int_1^{+\infty} x^{-3/4} . \cos\left[\ln(x) . \frac{t}{2}\right] . (\theta(x)-1) . dx$$

$$\xi(1/2 + i.t) = \xi(1/2 - i.t)$$

$$\xi\left(\frac{1}{2}+it\right) = \frac{1}{2}\left[\frac{1}{2}+it\right]\left[-\frac{1}{2}+it\right]\Gamma\left(\frac{1}{4}+\frac{it}{2}\right)\pi^{-\frac{1}{4}-\frac{it}{2}}\zeta\left(\frac{1}{2}+it\right)$$

$$= -\frac{1}{2}\exp\left[\operatorname{Re}\log\Gamma\left(\frac{1}{4}+\frac{it}{2}\right)\right]\pi^{-\frac{1}{4}}\left(t^2+\frac{1}{4}\right)Z(t),$$

Riemann-Siegel Zeta-function $\theta(0)=0$ $Z(0)=\zeta(1/2)= -1.46$

$$Z(t) = e^{i\vartheta(t)}\zeta\left(\frac{1}{2}+it\right) = \exp\left[i\operatorname{Im}\log\Gamma\left(\frac{1}{4}+\frac{it}{2}\right)-i\frac{\log\pi}{2}t\right]\zeta\left(\frac{1}{2}+it\right)$$

$$\exp\left[\operatorname{Re}\,\log\,\Gamma(\frac{1}{4} + \frac{i.t}{2})\right] = \exp\left[\operatorname{Re}\,\log\,\Gamma(\frac{1}{4} - \frac{i.t}{2})\right]$$

$$Z(t) = e^{i.\theta(t)} . \zeta(\frac{1}{2} + i.t) = e^{-i.\theta(t)} . \zeta(\frac{1}{2} - i.t)$$

Summary of the Riemann Hypothesis

$$\theta(x) = \sum_{n=-\infty}^{+\infty} e^{-n^2 . \pi . x} \qquad \frac{\theta(x)-1}{2} = \sum_{n=1}^{+\infty} e^{-n^2 . \pi . x} \qquad \frac{\theta(x)}{\theta(1/x)} = \frac{1}{\sqrt{x}}$$

$$X(s) = \int_0^{+\infty} x^{s/2 - 1} . (\frac{\theta(x)-1}{2}) . dx = \pi^{-s/2} . \Gamma(\frac{s}{2}) . \zeta(s)$$

$$X(s) = \frac{-1}{s.(1-s)} + \int_1^{+\infty} \left[x^{s/2 - 1} + x^{(1-s)/2 - 1} \right] . (\frac{\theta(x)-1}{2}) . dx$$

$$\text{Real } s > 1$$

$X(s)=X(1-s)$ **Analytic continuation**

See: Understanding the Zeta function

$$\xi(s) = \frac{-s.(1-s)}{2} . X(s) = \frac{-s.(1-s)}{2} . \pi^{-s/2} . \Gamma(\frac{s}{2}) . \zeta(s)$$

$$\xi(s) = \frac{1}{2} - \frac{s.(1-s)}{2} . \int_1^{+\infty} \left[x^{s/2 - 1} + x^{(1-s)/2 - 1} \right] . (\frac{\theta(x)-1}{2}) . dx$$

$$\xi(\frac{1}{2} + i.t) = \frac{1}{2} - \frac{(1/4)+t^2}{2} . \int_1^{+\infty} x^{-3/4} . \cos\left[\ln(x) . \frac{t}{2} \right] . (\theta(x)-1) . dx$$

$$\xi(1/2 + i.t) = \xi(1/2 - i.t)$$

$$\xi\left(\frac{1}{2}+it\right) = \frac{1}{2}\left[\frac{1}{2}+it\right]\left[-\frac{1}{2}+it\right]\Gamma\left(\frac{1}{4}+\frac{it}{2}\right)\pi^{-\frac{1}{4}-\frac{it}{2}}\zeta\left(\frac{1}{2}+it\right)$$

$$= -\frac{1}{2}\exp\left[\operatorname{Re}\log\Gamma\left(\frac{1}{4}+\frac{it}{2}\right)\right]\pi^{-\frac{1}{4}}\left(t^2+\frac{1}{4}\right)Z(t),$$

$\theta(0)=0 \quad Z(0)=\zeta(1/2)= -1.46$

$$Z(t) = e^{i\theta(t)}\zeta\left(\frac{1}{2}+it\right) = \exp\left[i\operatorname{Im}\log\Gamma\left(\frac{1}{4}+\frac{it}{2}\right)-i\frac{\log\pi}{2}t\right]\zeta\left(\frac{1}{2}+it\right)$$

Z(t) is the Riemann Siegel function

$$\vartheta(t) \sim \frac{t}{2}\log\frac{t}{2\pi} - \frac{t}{2} - \frac{\pi}{8} + \frac{1}{48t}$$

$$\exp\left[\operatorname{Re} \log \Gamma(-\frac{1}{4} + \frac{i.t}{2}) \right] = \exp\left[\operatorname{Re} \log \Gamma(-\frac{1}{4} - \frac{i.t}{2}) \right]$$

$$Z(t) = e^{i.\theta(t)} . \zeta(\frac{1}{2} + i.t) = e^{-i.\theta(t)} . \zeta(\frac{1}{2} - i.t)$$

Graph of Z(t) in the range of t (0⩽ t ⩽50)

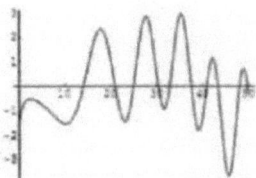

$$Z(t) = e^{i\theta(t)} \sum_{n=1}^{[x]} \frac{1}{n^{\frac{1}{2}+it}} + e^{-i\theta(t)} \sum_{n=1}^{[x]} \frac{1}{n^{\frac{1}{2}-it}} + O(t^{-\frac{1}{4}})$$

$$= 2 \sum_{n=1}^{[x]} \frac{\cos(\theta(t) - t\log n)}{n^{\frac{1}{2}}} + O(t^{-\frac{1}{4}}).$$

$$[x]=m$$

$$\tau = \sqrt{\frac{t}{2\pi}}, \quad m = \lfloor \tau \rfloor, \quad z = 2(t-m) - 1$$

**The estimation of the remaining terms is a science by itself
exclusively for the members of the top club
of Prime Numbers Theorem**

$$Z(t) = 2 \sum_{n=1}^{m} \frac{\cos(\theta(t) - t\log n)}{\sqrt{n}} +$$

$$(-1)^{m+1}\tau^{-1/2} \sum_{j=0}^{M} (-1)^j \tau^{-j} \Phi_j(z) + R_M(t)$$

The first functions $\Phi_j(z)$ are defined by

$$\Phi_0(z) = \frac{\cos(\frac{1}{2}\pi z^2 + \frac{3}{8}\pi)}{\cos(\pi z)}$$

$$\Phi_1(z) = \frac{1}{12\pi^2}\Phi_0^{(3)}(z)$$

$$\Phi_2(z) = \frac{1}{16\pi^2}\Phi_0^{(2)}(z) + \frac{1}{288\pi^4}\Phi_0^{(6)}(z)$$

The general expression of $\Phi_j(z)$ for $j > 2$ is quite complicated and we refer to [6]
or [10] for it. As exposed in [2], explicit bounds have been rigorously obtained
on the error term $R_M(t)$, and for $t \geq 200$, one has

$$|R_0(t)| \leq 0.127\,t^{-3/4}, \quad |R_1(t)| \leq 0.053\,t^{-5/4}, \quad |R_2(t)| \leq 0.011\,t^{-7/4}.$$

Numerical evaluation of the Riemann Zeta function
Xavier Gourdon and Pascal Sebah
July 23, 2003[1]

To be able to completly approximate $Z(t)$ thanks to formula (9), it remains to give an approximation of the $\theta(t)$ function which is obtained from expression (6) using Stirling series, giving

$$\theta(t) = \frac{t}{2} \log \frac{t}{2\pi} - \frac{t}{2} - \frac{\pi}{8} + \frac{1}{48t} + \frac{7}{5760t^3} + \cdots \qquad (10)$$

Practical approximation considerations

For practical purposes in computations relative to zeros of $\zeta(s)$ it is not necessary to compute precisely the zeros but just to locate them, and using $M = 1$ or $M = 2$ in formula (9) is usually sufficient. For t around 10^{10} for example, the choice $M = 1$ permits to obtain an absolute precision of $Z(t)$ smaller than 2×10^{-14}, and with $M = 2$ the precision is smaller than 5×10^{-20}. As for the

number of terms involved in the sum of (9), it is proportional to $\sqrt{t}$ which is much better than previous approaches without Riemann-Siegel formula which required a number of terms of order t. For $t \simeq 10^{10}$ for example, Riemann-Siegel formula only requires $m \simeq 40,000$ terms, whereas approach of proposition 1 requires at least $\simeq 9 \times 10^9$ terms.

As an Epilogue

Robin Gandy used to say, when working on something it is like moving from one dark room to another, trying to illuminate it, until finally you break through and come out to full sunlight!

As I mentioned my purpose was to get a satisfactory over all idea of the Zeta function.
I believe that this is achieved
although certain proofs were not rigorously approached

In their pdf Xavier Gourdon & Pascal Sebah provide some explanations to the derivation of the Riemann Siegel formula.
The remaining terms of the above formula require high level complex analysis.

John Bredakis MD

With the aid of the pdf by Gourdon – Sebah , I believe that

I made another significant step in understating the marvellous Zeta function, not

only for me but also for anyone with the basic knowledge of the gamma function.

And all this in relatively few pages

Most of the pdf in the internet related to the Zeta function are obviously adressed

exclusively to the other members of the club of Prime Numbers Theorem, to the point

of being incomprehensible to the mazority of those outside the club.

I like to express my gratitude to Theodore Yoder , for his pdf on Riemann Hypothesis

The pdf by T.Yoder was the only one , with very few elements of complex analysis.

The key idea of analytic continuation Iv got also from Theodore Yoder's pdf

_An Introduction to the **Riemann Hypothesis**
Theodore J. Yoder* May 27, 2011
Abstract Here we discuss the
most famous unsolved problem in mathematics,

I relied mainly on the following pdf in the Internet

The Wikipedia , the free encyclopedia

An **Introduction** to the **Riemann Hypothesis**
An Introduction to the Riemann Hypothesis Theodore J. Yoder* May 27, 2011 Abstract
Here we discuss the most famous unsolved problem in mathematics,
sections.maa.org/epadel/students/studentWinners/2011_Yoder.p ...

Zeros of the Riemann Zeta-function **on the critical line**
Universit a degli Studi ROMA TRE Zeros of the Riemann Zeta-function on the critical line
Author:
Lorenzo Menici Supervisor: Prof. Francesco Pappalardi
www.mat.uniroma3.it/scuola_orientamento/alumni/laureati/meni ...

Riemann Zeta Function - ONID
Riemann Zeta Function Bent E. Petersen January 23, 1996 ... the students learned about the
Riemann hypothesis { an important part of our mathematical heritage and culture.
people.oregonstate.edu/~peterseb/misc/docs/zeta.pdf

Numerical evaluation of the Riemann Zeta-function
PDF: Xavier Gourdon and Pascal Sebah
1. *Numerical evaluation of the Riemann. Zeta-function. Xavier Gourdon and Pascal* Sebah. July
23, 20031. We expose techniques that permit to approximate the ..

Odlyzko–Schönhage **algorithm** - Wikipedia, the free encyclopedia
In mathematics, the Odlyzko–Schönhage algorithm is a fast algorithm for evaluating the
Riemann
zeta function at many points, introduced by (Odlyzko & Schönhage 1988).
en.wikipedia.org/wiki/Odlyzko-Schonhage

And my pdf :
Big Bang in Math , abstract form of my method , **John Bredakis method in brief**
The Gamma function is difficult to proove and easy to use under a proper guidance.
http// Mathhighways.blogspot.com / John Bredakis

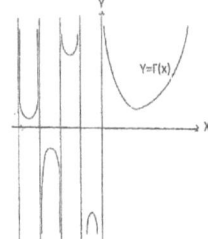
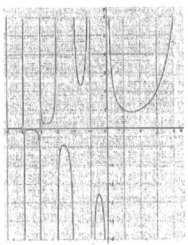

A Summary of the Gamma function $\Gamma(x)$

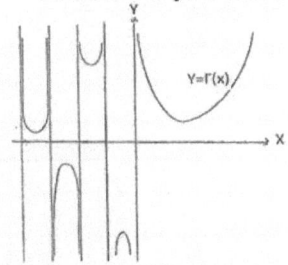
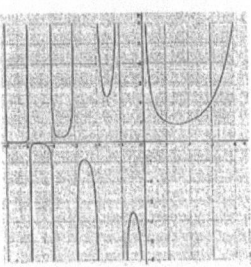

$$\Gamma(x) = \int_{0}^{+\infty} e^{-t} . t^{x-1} . dt = \lim_{n \to +\infty} \frac{n! . n^{x}}{x . (x+1) . (x+2) . \ \dots \ . (x+n)} \qquad \begin{array}{l} x \# 0 \\ x \# -1 \\ x \# -2 \\ \text{etc} \end{array}$$

Main property: $x . \Gamma(x) = \Gamma(x+1)$

$$\Gamma(x) = \frac{\Gamma(x+1)}{x} = \frac{\Gamma(x+2)}{x . (x+1)} = \frac{\Gamma(x+3)}{x . (x+1) . (x+2)} \text{etc} \ \left| \ \frac{\Gamma(x)}{\Gamma(n+x+1)} = \frac{1}{x . (x+1) .. (x+n)} \right.$$

$$\Gamma(n+1) = n! \ \left| \ \Gamma(n+\tfrac{1}{2}) = \frac{(2n)!}{2^{2n} . n!} . \sqrt{\pi} = \frac{(2n)!}{2^{n} . n!} . \frac{1}{2^{n}} . \sqrt{\pi} \right.$$

$\Gamma(1) = 0! = 1 \qquad \Gamma(2) = 1! = 1 \qquad \Gamma(3) = 2! = 2 \qquad \Gamma(4) = 3! = 6 \qquad \Gamma(5) = 4! = 24$ etc

$$\Gamma(\tfrac{1}{2}) = \sqrt{\pi} \ \left| \ \Gamma(\tfrac{3}{2}) = \tfrac{1}{2} . \sqrt{\pi} \ \right| \ \Gamma(\tfrac{5}{2}) = \tfrac{3}{2} . \tfrac{1}{2} . \sqrt{\pi} \ \left| \ \Gamma(\tfrac{7}{2}) = \tfrac{5}{2} . \tfrac{3}{2} . \tfrac{1}{2} . \sqrt{\pi} \ \right| \ \Gamma(\tfrac{9}{2}) = \tfrac{7}{2} . \tfrac{5}{2} . \tfrac{3}{2} . \tfrac{1}{2} . \sqrt{\pi} \ \text{etc}$$

$$\int_{0}^{+\infty} e^{-s.t} . t^{x-1} . dt = 2. \int_{0}^{+\infty} e^{-s.t^{2}} . t^{2x-1} . dt = \int_{0}^{1} t^{s-1} . \left[\ln(\tfrac{1}{t}) \right]^{x-1} . dt = \frac{\Gamma(x)}{x^{s}}$$

$$s > 0 \ \ x \# 0, -1, -2, \text{etc} \qquad s > 0 \quad x > 0 \qquad s > 0 \ \ s \# x \quad x > 0$$

Valid also for s=1

Legendre's duplication formula $\qquad \Gamma(2x) = \dfrac{1}{\sqrt{\pi}} . 2^{2x-1} . \Gamma(x) . \Gamma(x+\tfrac{1}{2})$
$\qquad\qquad\qquad x > 0$

$$\Gamma(x) . \Gamma(1-x) = \frac{\pi}{\sin(\pi.x)} \ \left| \ \Gamma(x) . \Gamma(-x) = \frac{\pi}{-x . \sin(\pi.x)} \ \right| \ \Gamma(\tfrac{1}{2}+x) . \Gamma(\tfrac{1}{2}-x) = \frac{\pi}{\cos(\pi.x)}$$

Denominator # 0

π **stands for products**

$$\frac{1}{\Gamma(x)} = \lim_{n\to+\infty} \frac{x.(x+1).(x+2)..(x+n)}{n!.n^x} = \frac{x.\prod_{n=1}^{+\infty}\left[1+\frac{x}{n}\right]}{n^x}$$

$$x \neq 0,-1,-2,-3,-4 \text{ etc} = e^{\gamma.x}.x.\prod_{n=1}^{+\infty}\left[1+\frac{x}{n}\right].e^{-x/n}$$

$$e^{\gamma.x} = e^{\left[1+\frac{1}{2}+\frac{1}{3}+\frac{1}{4}+..+\frac{1}{n}-\ln n\right].x} = \frac{\prod_{n=1}^{+\infty}\left[e^{x/n}\right]}{n^x}$$

$$n\to+\infty$$

$$\gamma = \lim_{n\to+\infty}\left[1+\frac{1}{2}+\frac{1}{3}+..+\frac{1}{n}-\ln n\right] = 0.577216 = \textbf{Euler's constant}$$

$$\ln\left[\frac{1}{\Gamma(x)}\right] = \gamma.x + \ln x + \sum_{n=1}^{+\infty}\ln\left[1+\frac{x}{n}\right] + \sum_{n=1}^{+\infty}\ln\left[e^{-x/n}\right]$$

$$\frac{d}{dx}\ln\left[\frac{1}{\Gamma(x)}\right] = -\frac{d}{dx}\ln\Gamma(x) = -\frac{\Gamma'(x)}{\Gamma(x)} = \gamma + \frac{1}{x} + \sum_{n=1}^{+\infty}\left[\frac{1}{n+x}\right] - \sum_{n=1}^{+\infty}\left[\frac{1}{n}\right]$$

$$\text{Or}\quad \frac{\Gamma'(x)}{\Gamma(x)} = \sum_{n=1}^{+\infty}\left[\frac{1}{n}-\frac{1}{n+x}\right] - \gamma - \frac{1}{x}$$

$$\frac{\Gamma'(k+1)}{\Gamma(k+1)} = 1 + \frac{1}{2} + \frac{1}{3} + \frac{1}{4} + \frac{1}{5} + \frac{1}{6} + \frac{1}{7} +..+ \frac{1}{k} - \gamma = \left[\phi(k)-\gamma\right]$$

k=a positive integer or zero $\phi(0)=0$ **by convention**

$$\frac{\Gamma'(1)}{\Gamma(1)} = -\gamma \quad\bigg|\quad \frac{\Gamma'(2)}{\Gamma(2)} = 1-\gamma \quad\bigg|\quad \frac{\Gamma'(3)}{\Gamma(3)} = 1+\frac{1}{2}-\gamma \quad\bigg|\quad \frac{\Gamma'(4)}{\Gamma(4)} = 1+\frac{1}{2}+\frac{1}{3}-\gamma$$

$$n=(0,1,2,3,\ldots,n) \qquad\qquad n!=\Gamma(n+1)$$

$$\int_0^{+\infty} e^{-s.t}.t^n.dt = 2.\int_0^{+\infty} e^{-s.t^2}.t^{2n+1}.dt = \int_0^1 t^{s-1}.\left[\ln\left(\frac{1}{t}\right)\right]^n.dt = \frac{\Gamma(n+1)}{s^{n+1}}$$
$$s>0 \qquad\qquad s>0 \qquad\qquad s>0$$

$$\Gamma(n+1)=n! \quad\Bigg|\quad \Gamma\left(n+\frac{1}{2}\right) = \frac{(2n)!}{2^{2n}.n!}.\sqrt{\pi} = \frac{(2n)!}{2^n.n!}.\frac{1}{2^n}.\sqrt{\pi}$$

$$\Gamma(1)=0!=1 \qquad \Gamma(2)=1!=1 \qquad \Gamma(3)=2!=2 \qquad \Gamma(4)=3!=6 \qquad \Gamma(5)=4!=24 \text{ etc}$$

$$\Gamma\left(\frac{1}{2}\right)=\sqrt{\pi} \,\Bigg|\, \Gamma\left(\frac{3}{2}\right)=\frac{1}{2}.\sqrt{\pi} \,\Bigg|\, \Gamma\left(\frac{5}{2}\right)=\frac{3}{2}.\frac{1}{2}.\sqrt{\pi} \,\Bigg|\, \Gamma\left(\frac{7}{2}\right)=\frac{5}{2}.\frac{3}{2}.\frac{1}{2}.\sqrt{\pi} \,\Bigg|\, \Gamma\left(\frac{9}{2}\right)=\frac{7}{2}.\frac{5}{2}.\frac{3}{2}.\frac{1}{2}.\sqrt{\pi} \text{ etc}$$

$$n=\text{any non negative integer}=(0,1,2,3,4,\ldots,n)$$

$$\int_0^{+\infty} e^{-s.t}.t^{n-(1/2)}.dt = 2.\int_0^{+\infty} e^{-s.t^2}.t^{2n}.dt \qquad = \frac{\Gamma\left(n+\frac{1}{2}\right)}{s^{\left(n+\frac{1}{2}\right)}}$$
$$s>0$$
$$= \int_0^1 t^{s-1}.\left[\ln\left(\frac{1}{t}\right)\right]^{n-(1/2)}.dt$$

The Beta function $B(x,y)$

$$B(x,y) = \frac{\Gamma(x).\Gamma(y)}{\Gamma(x+y)} \qquad \boxed{x,y>0}$$

$$B(x,y) = \int_0^1 (1-t)^{y-1}.t^{x-1}.dt = \int_0^{+\infty} \frac{u^{x-1}}{(1+u)^{x+y}}.du = 2.\int_0^{\pi/2} \sin^{2x-1}\theta.\cos^{2y-1}\theta.d\theta$$

The only way to remember the formulas of the gamma function is through repeated use , and a good source like my pdf where those formulas are clearly written.

Numerical table of Γ(x)

x	Γ(x)	x	Γ(x)	x	Γ(x)	x	Γ(x)
1.0	1.00000	1.25	0.90840	1.50	0.88623	1.75	0.91906
.01	0.99433	.26	0.90440	.51	0.88659	.76	0.92137
.02	0.98884	.27	0.90250	.52	0.88704	.77	0.92376
.03	0.98355	.28	0.90072	.53	0.88757	.78	0.92623
.04	0.97844	.29	0.89904	.54	0.88818	.79	0.92877
.05	0.97350	1.30	0.89747	1.55	0.88887	1.80	0.93138
.06	0.96874	.31	0.89600	.56	0.88964	.81	0.93408
.07	0.96415	.32	0.89600	.57	0.89049	.82	0.93685
.08	0.95973	.33	0.89464	.58	0.89142	.83	0.93969
.09	0.95546	.34	0.89222	.59	0.89243	.84	0.94261
1.10	0.95135	1.35	0.89115	1.60	0.89352	1.85	0.94561
.11	0.94740	.36	0.89018	.61	0.89468	.86	0.94869
.12	0.94359	.37	0.88931	.62	0.89592	.87	0.95184
.13	0.93993	.38	0.88854	.63	0.89724	.88	0.95507
.14	0.93642	.39	0.88785	.64	0.89864	.89	0.95838
1.15	0.93304	1.40	0.88726	1.65	0.90012	1.90	0.96177
.16	0.92980	.41	0.88676	.66	0.90167	.91	0.96523
.17	0.92670	.42	0.88636	.67	0.90330	.92	0.96877
.18	0.92373	.43	0.88604	.68	0.90500	.93	0.97240
.19	0.92089	.44	0.88581	.69	0.90678	.94	0.97610
1.20	0.91817	1.45	0.88566	1.70	0.90864	1.95	0.97988
.21	0.91558	.46	0.88560	.71	0.91057	.96	0.98374
.22	0.91311	.47	0.88563	.72	0.91258	.97	0.98768
.23	0.91075	.48	0.88575	.73	0.91467	.98	0.99171
.24	0.90852	.49	0.88595	.74	0.91683	.99	0.99581
						2.0	1.00000

Main property: $x.\Gamma(x)=\Gamma(x+1)$

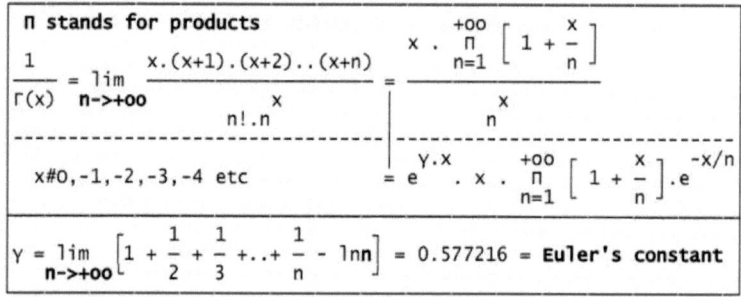

$$\Pi \text{ stands for products}$$

$$\frac{1}{\Gamma(x)} = \lim_{n\to+\infty} \frac{x.(x+1).(x+2)..(x+n)}{n!.n^x} = \frac{x . \prod_{n=1}^{+\infty} \left[1 + \frac{x}{n} \right]}{x^n}$$

$$x \neq 0,-1,-2,-3,-4 \text{ etc} \qquad = e^{\gamma.x} . x . \prod_{n=1}^{+\infty} \left[1 + \frac{x}{n} \right] . e^{-x/n}$$

$$\gamma = \lim_{n\to+\infty} \left[1 + \frac{1}{2} + \frac{1}{3} +..+ \frac{1}{n} - \ln n \right] = 0.577216 = \textbf{Euler's constant}$$

$$= 0.577215665$$

Contribution of the Gamma function $\Gamma(x)$
to the normal distribution

$\int_{0}^{+\infty} e^{-s.t^2}.dt = \dfrac{\Gamma(1/2)}{2.s^{1/2}}$ $s>0$	$\int_{0}^{+\infty} e^{-s.t^2}.t.dt = \dfrac{\Gamma(1)}{2.s}$ $s>0$	$\int_{0}^{+\infty} e^{-s.t^2}.t^2.dt = \dfrac{\Gamma(3/2)}{2.s^{3/2}}$ $s>0$

Even   **Odd**   **Even**

Integrand functions

$\Gamma(3/2)=(1/2).\Gamma(1/2)$

$\int_{-\infty}^{+\infty} \sqrt{\dfrac{s}{\pi}}.e^{-s.t^2}.dt = 1$ $s>0$	$\int_{-\infty}^{+\infty} \sqrt{\dfrac{s}{\pi}}.e^{-s.t^2}.t.dt = 0$ $s>0$	$\int_{-\infty}^{+\infty} \sqrt{\dfrac{s}{\pi}}.e^{-s.t^2}.t^2.dt = \dfrac{1}{2.s}$ $s>0$

Setting instead of $\sqrt{s}$ **the** $\dfrac{1}{\sigma.\sqrt{2}}$ $\sigma>0$ **and** $(x-xo)$ **instead of** t **we get:**

Total probability	Mathematical expectation	Variance
$\int_{-\infty}^{+\infty} \phi(x).dx = 1$	$\int_{-\infty}^{+\infty} \phi(x).x.dx = xo$	$\int_{-\infty}^{+\infty} \phi(x).(x-xo)^2.dx = \sigma^2$

$\phi(t) = \dfrac{1}{\sigma.\sqrt{2\pi}}.e^{-\frac{1}{2}.\left(\frac{t}{\sigma}\right)^2}$	$\phi(x) = \dfrac{1}{\sigma.\sqrt{2\pi}}.e^{-\frac{1}{2}.\left(\frac{x-xo}{\sigma}\right)^2}$ **By shifting of axis t**

Non Conditional Probability
Random Variable $X=\{xn\}$ - **Random Numbers** $\{xn\}=\{x1,x2,..,xn\}$

$\Phi(x)$=Probability distribution $\phi(x)$=Probability density	$P(a\backslash<X<b)=\Phi(x)\Big|_a^b = \int_{a}^{b} \phi(x).dx$

. The integral is replaced by the sum for discrete random variables

. The $\Phi(x)$ for continuous random variables is the area under the probability density function $\phi(x)$.

. We can start with values of X within the range of reality or from -oo to +oo considering the $\phi(x)$ outside the range of reality as zero.

Applications of the Beta function B(x,y)

$$B(x,y) \quad B(x,y) = \frac{\Gamma(x).\Gamma(y)}{\Gamma(x+y)} \quad \boxed{x,y > 0}$$

$$= \int_0^1 (1-t)^{y-1}.t^{x-1}.dt = \int_0^{+\infty} \frac{u^{x-1}}{(1+u)^{x+y}}.du = 2. \int_0^{\pi/2} \sin^{2x-1}\theta.\cos^{2y-1}\theta.d\theta$$

k=a positive integer ■ $x^k = t \Rightarrow t^z = x$ ■ $z=(1/k)$

$$\int_0^{+\infty} \frac{dx}{1+x^k} = \int_0^{+\infty} \frac{dt^{1/k}}{1+t} = \frac{1}{k}.\int_0^{+\infty} \frac{t^{z-1}}{1+t}.dt \overset{z=(1/k)}{=} \frac{1}{k}.\frac{\Gamma(z).\Gamma(1-z)}{\Gamma(1)} = \frac{1}{k}.\frac{\pi}{\sin(\pi.z)}$$

$$\int_0^{+\infty} \frac{dx}{1+x^4} = \frac{1}{4}.\int_0^{+\infty} \frac{t^{z-1}}{1+t}.dt \overset{z=(1/4)}{=} \frac{1}{4}.\frac{\Gamma(1/4).\Gamma(3/4)}{\Gamma(1)} = \frac{1}{4}.\pi.\sqrt{2}$$

$$\int_0^1 \frac{dx}{\sqrt[k]{1-x}} = \frac{1}{k}.\int_0^1 (1-t)^{-(1/2)}.t^{z-1}.dt \overset{\boxed{z=(1/k)}}{=} \frac{1}{k}.\frac{\Gamma(1/2).\Gamma(z)}{\Gamma(\frac{1}{k}+\frac{1}{2})}$$

$$\int_0^1 \frac{dx}{\sqrt[3]{1-x}} = \frac{1}{3}.\int_0^1 (1-t)^{-(1/2)}.t^{z-1}.dt \overset{\boxed{z=(1/3)}}{=} \frac{1}{3}.\frac{\Gamma(1/2).\Gamma(1/3)}{\Gamma(\frac{1}{3}+\frac{1}{2})}$$

$$\int_0^1 t^m.(1-t)^n.dt = B(m+1,n+1) = \frac{\Gamma(m+1).\Gamma(n+1)}{\Gamma(m+n+2)} = \frac{m!.n!}{(m+n+1)!}$$

$$\int_0^\pi \sin^{2x-1}\theta.d\theta = 2.\int_0^{\pi/2} \sin^{2x-1}\theta.d\theta = B(x,\frac{1}{2}) = \frac{\Gamma(x).\Gamma(1/2)}{\Gamma(x+1/2)} \quad \boxed{x>0}$$

Additional References:

1.   **Higher Mathematics for beginners:**
       ----
                by Ya.B.Zeldovich
            (Mir Publishers Moscow 1973)

2.   **Calculus with analytic geometry:**
       ----
       by Harley Flanders and Justin J Price
                (Academic Press 1978)

3.   **A brief course of higher mathematics:**
       ----
       by V.A.Kudryavtsev and B.P.Demidovich
             (Mir Publisher's Moscow 1980)

4.   **Concice Encyclopedia of Mathematics:**
       ----
        by W.Gellert,H.Kustner,M Hellwich,H Kastner
    (Van Nostrand Reinhold Company New York and other cities 1977)

5.   **Computational Mathematics:**
       ----
        by B.P Demitovich and I.A.Maron
            (Mir publishers Moscow 1976)

6.   **Advanced calculus:**
       ----
               by Leopold Flatto
    (The wiiliams and Wilkins Company - Baltimore 1982)

7. **Mathematics Handbook for Science and Engineering:**
    ----
                            ^
       by: Royal Lennart Rade and Bertil Westegren
                    Fifth edition - 2004
                Springer Verlag Publications Inc
                 Berlin - Heidelberg - New York

8. **Mathematical methods for physicists and engineers:**
    ----
        by: Royal Eugene Collins - 2nd corrected edition
       Dover Publications Inc - Mineola New York - USA 1991

9.   **Differential Equations:**

      A systems approach - by: Jack Goldberg - and Merle C.Potter
              Prentice Hall International Editions
              Upper Saddle River , NJ - USA - 1998

- And a lot of personal work -

John.K.Bredakis MD

Assistant Professor University of Athens

American Board Certified Cardiologist

Born in Athens Greece 28/11/1946

**Graduate of the medical school (1970)
University of Athens**

**Trained in internal medicine and Cardiology
(1970-1977)
Chicago - USA**

**Consultant Cardiologist - Areteion Hospital Athens Greece
Since 1977**

Thanks God , uncle Fotis , Areteion Hospital
my parents , my wife Sofia
and professors C.Tountas , D.Voros , G.Limouris

A special thanks also to the professor Elias Kastanas
(Professor of mathematics - Engineer - Computer scientist etc)
The creator of my blog

I would also like to thank very much the professor of mathematics
Themistoklis Rassias

Athens Greece 2012